SESAME STREET®

A Trip to the ZOO with SESAME STREET

Christy Peterson

Lerner Publications ◆ Minneapolis

Elmo and his friends from *Sesame Street* are going on a field trip, and you're invited! Field trips provide children with the opportunity to explore their communities, visit new places, and experience hands-on learning. This series brings the joys of field trips to your fingertips. Where will you go next?

—Sincerely, the Editors at Sesame Street

TABLE OF CONTENTS

LET'S VISIT THE ZOO!

Today we are visiting the zoo. Zoos teach people about animals.

5

Animals live in habitats. A habitat is a place where animals can find food, water, and somewhere to sleep. This habitat has cold water and rocks for the penguins.

The best habitat for me is in my trash can with lots of junk!

I ♥ Trash

This habitat has branches and ropes. Can you guess who lives here?

Chimpanzees live here! Their habitat looks like a playground.

Zookeepers take care of the animals. They clean the habitats. They give each animal healthy foods.

This tiger cub is drinking milk. It sure look tasty!

Animals are all different shapes and sizes. This helps each animal find the food it needs. For example, a giraffe has a long neck to reach the leaves of tall trees.

Do you see the giraffe's long tongue? The giraffe uses its tongue to pull the leaves off the tree.

Zookeepers know lots of things about all the animals. A zookeeper tells us that this tortoise can live to be about 150 years old.

The veterinarian at the zoo helps keep the animals healthy. The veterinarian cares for them when they are hurt or sick.

A veterinarian gives each animal a checkup, just like I get at the doctor!

Volunteers at the zoo help teach people about animals.

They answer questions and take groups on tours.

Zoos help us learn all about different animals from around the world! What's your favorite animal to see at the zoo?

Elmo had so much fun at the zoo! Elmo loved seeing the elephants!

THE ZOO AT HOME

You can make a bird feeder out of a clean, dry milk carton. You'll need the carton, some string, sturdy scissors, birdseed, and an adult's help.

- Have an adult poke two holes in the top of the carton and push the string through the holes.

- Ask an adult to then cut out two opposite sides of the carton. Leave a little of the side near the bottom so the seed won't fall out.

- Put birdseed in the bottom of the feeder.

- Then use the string to hang the feeder in your yard or on a balcony.

GLOSSARY

habitat: a place where animals can find food, water, and somewhere to sleep

veterinarian: a doctor that takes care of animals and helps keep them healthy

volunteer: someone who does something without pay

zookeeper: a person who takes care of zoo animals

LEARN MORE

Carson, Mary Kay. *Wildlife Ranger Action Guide.* North Adams, MA: Storey, 2020.

Clark, Rosalyn. *A Visit to the Zoo.* Minneapolis: Lerner Publications, 2018.

Salzmann, Mary Elizabeth. *Zoo Babies.* Minneapolis: Sandcastle, 2020.

INDEX

PHOTO ACKNOWLEDGMENTS

Image credits: Torychemistry/Shutterstock.com, p. 5; Eugenia P/Shutterstock.com, p. 6; Pontus Edenberg/Shutterstock.com, p. 8; Grant Faint/The Image Bank/Getty Images, p. 9; ton koene/Alamy Stock Photo, p. 10; Dimas Ardian/Getty Images, p. 11; davegkugler/Shutterstock.com, p. 12; nik wheeler/Alamy Stock Photo, p. 13; Anthony Devlin/EMPPL PA Wire via AP Images, p. 14; Deni Williams/Shutterstock.com, p. 16; Vietnam stock photos/Shutterstock.com, p. 18; kali9/E+/Getty Images, p. 19; Tatiana Litvinova/Shutterstock.com, p. 20.

Cover: Tetyana Kokhanets/Shutterstock.com.

Lerner Publications Company
An imprint of Lerner Publishing Group, Inc.
241 First Avenue North
Minneapolis, MN 55401 USA

For reading levels and more information, look up this title at www.lernerbooks.com.

Main body text set in Mikado a.
Typeface provided by HVD Fonts.

Editor: Andrea Nelson

Library of Congress Cataloging-in-Publication Data

Names: Peterson, Christy, author.
Title: A trip to the zoo with Sesame Street / Christy Peterson.
Description: Minneapolis : Lerner Publications, 2022. | Series: Sesame Street field trips | Includes bibliographical references and index. | Audience: Ages 4–8 | Audience: Grades K–1 | Summary: "Young readers explore the zoo with their Sesame Street friends to learn about animals, their habitats, and more. Readers also learn how to make a bird feeder from a milk carton." —Provided by publisher.
Identifiers: LCCN 2021010278 (print) | LCCN 2021010279 (ebook) | ISBN 9781728439136 (library binding) | ISBN 9781728445106 (ebook)
Subjects: LCSH: Zoos—Juvenile literature. | Zoo animals—Juvenile literature.
Classification: LCC QL76 .P48 2022 (print) | LCC QL76 (ebook) | DDC 590.73—dc23

LC record available at https://lccn.loc.gov/2021010278
LC ebook record available at https://lccn.loc.gov/2021010279

Manufactured in the United States of America
1-49819-49687-8/2/2021